ELIMINATE TWO PRINCIPAL CAUSES OF CHILD ILLNESS AND DEATH IN AFRICA: DIARRHEA AND MALARIA

A Peace Corps Writers Book — an imprint of Peace Corps Worldwide

Printed in the United States of America

by Peace Corps Writers of Oakland, California.

For more information, contact peacecorpsworldwide@gmail.com. Peace Corps Writers and the Peace Corps Writers colophon are trademarks of PeaceCorpsWorldwide.org.

ISBN 978-1-950444-51-9

Library of Congress Control Number: 2022921658

First Peace Corps Writers Edition, November 2022

Table of Contents

Dedication

This book is written to honor the many people I have had the privilege to work with and learn from and to help those who want to do impactful water projects. They include Peace Corps Volunteers and staff in Guatemala, particularly Gary Wederspahn, Frances Asturias, Tim Hardin, Jim Burchfield, Dennis McCarty, Pat Buckles, and others. My brothers Richard and Greg and my father always contributed to my efforts.

At the University of Georgia, Dr. Eugene Odum taught me the ecology of diseases and Dr. W. O. Caster provided me with experiments that showed me the importance of complementing grains and legumes with dark green leafy vegetables. At Hopkins, I had the privilege of studying with Drs. Cecile DeSweemer, Carl Taylor, Robert Parker, Bill Ward, Larry Green, and other knowledgeable professors.

As a contractor for the Office of International Health of the Department of Health, Education and Welfare, Dr. Naomi Baumslag inspired me and taught me the importance of breastfeeding, oral rehydration, and literature reviews, and Dr. Ed Sabin and his wife Robbie became life-long friends. Art French showed me additional concepts in surveying in our national survey of the USDA EFNEP program. Meeting Raj Arole and reading Raj and Bea's book on the Jamkhed Project taught me the importance of community grandmothers and their leadership and integrated programs, methods I have copied in many countries. Jim Borton in CARE Thailand and the World Food Programme enabled and complemented successful work in alleviating refugee and displaced person malnutrition and illness on the Thai-Cambodian Border. Mike Carroll, Ed Waters, Margie and Peter Morris have continued the journey with me. Dr. Rudi Horner was continually supportive and encouraging. In Bolivia, Frank Sullivan gave me the staff and many of the methods to understand the potential of water projects in development. Dr. Mario Andrade taught me excellent training dynamics and methods in Ecuador and has continue to be a friend ever since.

Working without pay, in Uganda, Ajaga Innocent Buran, Rukia Driciru, Dr. Brian Nzano, Lenny Christine, Moses Anguzu, Emmy Zoomlamai, Royce Gloria Androa, and many others have been companions in service and continual improvement. Helene Dudley of TCP Global and Jennifer Mally of Good Foundations International have been fabulous collaborators and inspirations. Dr. Melissa Masoner of the Topeka Rotary

Club has been a continual partner in efforts. Dr. Chris Bloore, Henk Holtslag, Jim McGill, Stephen Adkins, and Marcos Campos have been competent, thorough partners in developing Rotary Water, Sanitation, and Hygiene (WASH) criteria scales. In Rotary, Ron Denham, K. Neil Van Dine, Erica Gwynn, Heidi Rickels, and Steve Werner have welcomed me to WASH programming. Marvin Weisbord and Sandra Janoff taught me how to plan and conduct Future Search whole systems change conferences to overcome systems problems and wrote the guidebook I use.

My church community, Penn Valley Friends, particularly George and Deborah Black, supports service over self, encouraging my work and introducing me to partners like Drs. Ebby Luvaga and Imali Abala. Jeff Morgan of Morgan Miller Plumbers and Eric Lehan, civil engineer and professional member of Engineers without Borders, helped refine our field methods.

Also, friends from the distant and recent past have chipped in to help extend a hand to additional communities. They are over a hundred people.

Throughout all this work, my wife, Connie Benish, and my daughters, Martine and Renee, have supported my passion to collaborate and learn.

Thank you all.

Introduction

Billions of people live with too little and contaminated water, suffer hunger daily, have few resources, inadequate education, and little knowledge how to improve their lives. Frequently, the only law they know is humility: they should not speak and have no right to be heard. Unless this changes, they will remain without opportunity, not contributing, and suffering rather than living.

To give these people a hand up in life, we need to start by affirming and collaborating with their priorities, not ours.

Interviewing thousands, they say start with water. Safe water leads to good health, less work, and greater production. More water leads to cleanliness, increased crop production as much as twelve-fold, and more home and other industries. Joining our knowledge and creativity with theirs, their resources with ours, their initiative with ours, we can respond to the less privilegeds' needs and start them on the road to productive self-sufficiency, affirming and strengthening their competency and leadership abilities. We want them to be full participants in the world's creativity, production, and development.

Bringing the poor and ignored into self-sufficiency is a process. It starts with meeting them, listening to them, responding, and collaborating, and using best processes. What and how we do things is important. We can make the poor dependent or more self-sufficient. We can help them feel respected or disrespected. We can strengthen their knowledge of how to organize, how to plan together, how to assess opportunities and find relevant information, how to work together, and how to measure their successes, or we can weaken those competencies. Of course, we want to strengthen them.

In this book, we will share the optimal methods we learned during 50 years of experimenting with communities and partners and testing the best practices we could find. These processes give communities a hand up, not handouts. They empower.

What we share in the book is the importance of water. We also impart the importance of what we measure. The process is to start by consulting partners and communities, select the priorities of the poor, not ours, plan together, sign contracts that commit all of us to clear contributions, work progressively and continuously to improve the quality of life, consider possible add-ons, and learn from and improve based on each collaboration with communities and partners. Everyone needs to learn: us, the communities; partners; donors, and the government.

If we collaborate well, our work can be an enduring step on the path to progress and empowerment. If we do it poorly, it can be a step backwards and a waste of time, money, and effort.

We can continually improve what we do. Motion is a key sign of life. Let's make it forward motion.

Chapter 1. Why do Water, Sanitation, and Hygiene Projects?

Among hundreds of millions of people, two of the leading three causes of child disease, death, and expenditures, diarrhea and malaria, are water-related. About 90 diseases are water-related. Our surveys indicate that as much as 30% of household income on average is spent on treating diarrhea and malaria. For these reasons, millions of people want to benefit from water, sanitation, hygiene, and malaria control services. When services are constructed ideally and people improve their behaviors, these diseases and costs are prevented.

Additionally, obtaining water frequently requires hours of walking and carrying 42-pounds of water miles each day. When families do not have nearby water supplies, they frequently take their young daughters out of school to carry water and herd animals to and from water sources. This reality harms development: studies have shown mothers' education is reflected in the survival of their children.

Also, water is a productive resource. Industries such as cheese-making require good water supplies. Fish farming requires water. Raising livestock needs water. Vegetable gardening and farming result in as much as twelve-fold increases in production if there is a water supply.

Our objective needs to be to do development, to help the communities move from being fatalistic to controllers of their destinies to the extent possible. We want them to realize that their futures either are determined by someone else or that they can be active participants in determining them. We want them to decide to be active, not passive, to collect information, assess needs, and plan and carry out their plans, continually improving. That is the human condition lived optimally. There are many skills and understandings involved. We want to work with them, moving both them and ourselves to more beneficial achievement and skill levels. We start this process by asking what they want to work on and joining forces with them. In my experience, among many impoverished communities, one of the principal things they want is more and better water. Therefore, respecting them and their wishes, we start working on their priorities, validating their decisions, not ours. For millions of communities, the initial point of engagement is water. Ask them.

For all these reasons, supplying safe water and large quantities of water are highly desirable.

Despite these realities, when water, sanitation, and hygiene are not done properly, they may fail quickly, create new problems, or not have all the possible benefits. For instance, studies have shown that safe water alone, more water, sanitation, or hygiene alone may only decrease diarrheal diseases by 25%. We must design for the benefits we want: safety, quantity, sanitation, hygiene, productivity, and sustainability. We must improve the whole environment because many of the water-related

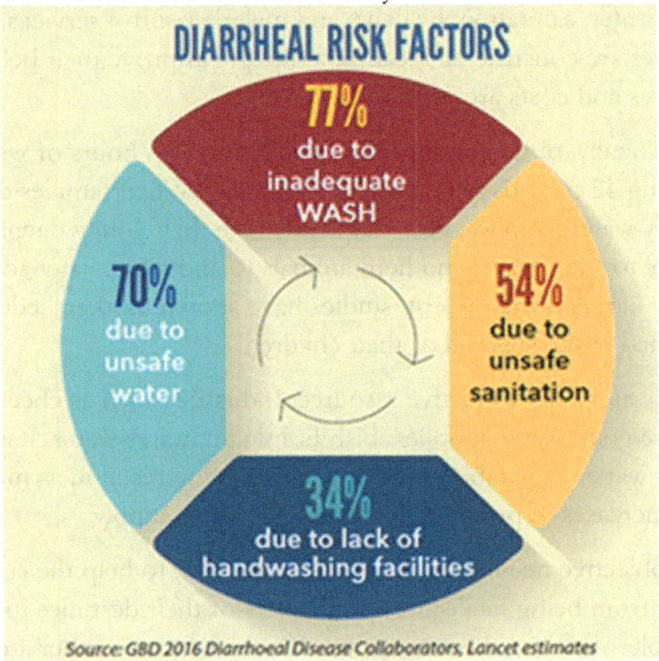

diseases are environmental, not individual.

In this book, we will share goals, measures, and some of the methods to achieve the maximum benefits from water, sanitation, and hygiene activities.

The following is Lancet's estimation of the impact of these different activities when they are done in isolated fashion.

On the other hand, when the water, sanitation, and hygiene (WASH) components are integrated, they result in disease rates like we have in the USA and other developed countries, practically eliminating diarrhea and malaria.

Chapter 2. Plan with the community

The objective of WASH projects is to improve the environment, so diseases and parasites are not transmitted, productivity and ease of work increase, expenses decrease, and the system endures. This is whole systems change and improvement.

There are methodologies that have been developed and tested that facilitate whole systems change. One that I have used many times is the Future Search Conference. It has books, a website, and an organization that helps those who want to use it. A link is Methodology | Future Search..

Summarizing the methodology, it follows the following steps:

1. Identify the systems problem to be addressed;
2. Scan the system as to whom are the key stakeholders (customers, providers, suppliers, CEO, workers, others affected, regulators, experts, and technicians);
3. Invite a representative of each stakeholder group to plan the future search;
4. Ask the stakeholders to identify whether anyone is missing who should be included in the planning group;
5. With the stakeholder group, identify a title, date, and place for the conference;
6. Count how many stakeholder groups there are;
7. Ask the stakeholder group planners to invite as many of their stakeholder group as there are stakeholder groups (e.g., if there are ten stakeholder groups, they should ask ten participants from each group, but the group planner is the one who invites them):
8. Request the invitees to commit to participate for the full length of the conference and to bring something that best represents what is most important to them in their lives;
9. Set up the conference in a comfortable site with round tables and chairs for all participant groups (e.g., ten round tables with ten chairs for each);
10. Briefly introduce the theme, objective, and method of the conference to the participants;

11. Ask each participant to stand and say who s/he is and show the object that represents him/her with a brief explanation what it shows that is so important;
12. Over the course of 16 hours, have participants share their recollections of the past, their analysis of the present, and what they would like the future to be in 20 years (follow the detailed process in a Future Search Conference book);
13. Record all suggestions and identify suggestions recommended across all groups (consensus futures);
14. Do action planning to achieve consensus futures by 1) all individuals write promises to themselves to be mailed back to them and 2) in voluntary planning, groups plan consensus futures;
15. Close the meeting, thanking participants, letting them share their feelings, and urging them to create the future they have imagined.

Chapter 3. Partner

Partnering with local Rotarians, indigenous non-profits, communities, and governmental departments has many benefits. Local organizations are permanent in the area, are very credible to communities, speak the local language and use local ideas, have influence with the communities and authorities, and can ensure the continuation of services.

In Uganda, we partner with local Rotary Clubs, the District, local non-profits, the communities, and the government. That makes our projects more impactful, lower cost, and sustainable.

Photos of some of our partners follow:

Rotary Club of Yumbe in front of CCEDUC offices. We partner with both.

Also, we partner with communities like the following:

Chapter 4. Zoom and plan with iNGO and local Rotary Club

To eliminate the costs of travel, hotels, meals, etc., we plan with communities and partners over the internet, using Zoom, WhatsApp, Go to Meeting, and other similar services.

Many planning and monitoring meetings can be seen on my YouTube channel at https://www.youtube.com/channel/UC8-AlPqoeZ7F22wH2VDMTEw

The most important aspects of these meetings are to introduce and recognize participants, have an agenda, facilitate participation, make and follow up action plans, adhere to time schedules, and summarize and share resolutions.

Chapter 5: Budget

A typical minimal budget for Rotary Clubs to do WASH in Uganda costs less than $20 per individual to bring them into living conditions that do not kill them.

For example, the following is one budget used in an actual project to serve about 60,000 people:

Item	Number	USD$
Bags of Cement	41000	$ 344,491.86
PVC tubes 6 m long	16000	$ 149,373.17
Pieces of fly screen	16000	$ 17,071.22
Reinforcing iron y 6 mm	16000	$ 104,561.22
Anti-malarials	75000	$ 25,406.78
Supplemental LLITNs and RDTs	15000	$ 85,278.74
Field staff	1	8,002.13
Field bike	1	$ 3,979.73
Boreholes, rehabilitated borehole, extra money to treadle pumps	100	$ 568,524.94
Computer	1	$ 1,137.05
Water filters	5000	$ 50,000.00
Cell phones with solar panel for recharging	102	$ 100,666.84
Cell phone connection fees	107	$ 3,424.91
Hero Bicycle for VHTs or Boda-boda expense where terrain does not allow	100	$ 9,335.82
Group personal accident insurance	102	$ 4,081.09
Gasoline	1	$ 5,879.97
Supplements to microloan groups	100	$ 400,000.00
Consultancies and travel		$ 10,000.00
Miscellaneous and inflation		$ 215,179.57
Total		$ 2,106,395.04

Chapter 6. Community follow-up assessment by Rotarians

Once a community has requested assistance and sent in its needs and intention to work to improve WASH and overcome malaria, it is important to visit and meet the people and their leadership personally. The objective is to verify their resolve and situation.

In Uganda, Rotary Clubs or indigenous non-profit counterparts meet with the communities to certify their commitments and needs. We leave the decision to collaborate to our partners, other than underscoring what our suggested criteria are and listening to their perspectives.

The following are photos of our partners visiting a new village that requested we work with them and the conditions they found.

Partners talking with villagers

Chapter 7. Survey

"You achieve what you measure," a director of curriculum for a school district told me.

What does that mean in WASH projects? Measure disease incidence before and after and compare it to other communities at the same point of time. Measure water quality and quantity. Measure walking time saved, money saved, and school attendance changed. If you expect the effect, measure it. Otherwise, you are relying on faith, belief, and assumptions.

Therefore, we designed surveys to be conducted at the start and after completing projects. Don't leave it to faith. Measure it.

We have developed and keep community surveys on https://www.mwater.co/. We can download these surveys to phones. Our partners implement them using our sampling instructions. The data uploads to the web and is analyzable in real time.

When designing a survey, various principles apply. One, sample, don't try to do every household. Two, use tested, standardized questions: don't design new ones that may not be comparable to those done nationally or internationally. Three, field test the questions: don't go in with questions that aren't understood.

Fortunately, mWater.co is a free internet service that allows the design, revisions, and deploy of surveys to smartphones anywhere. When the survey is done in the field, it uploads to the internet as soon as the phone is within transmission distance of a cell tower. Surveying can be done in extremely rural places, uploaded, tested, and analyzed that same day.

One example of a community pre- post-survey we use that is on mWater follows:

Community WASH survey

Administration

Which community are you surveying?

____

◯ Other (please specify)

Survey date for field work ________________________________

Characterize Survey Site ________________________________

Characterize area

- ◯ Urban
- ◯ Rural
- ◯ Transition between urban and rural

Name of head of household ________________________________

Hint: First Middle Last

Survey Location

Where is this survey site located? ________________________________

Consent and Photos

Do we have your permission to take photos of this person/house/latrine?

Hint: Explain why we are doing this survey and what we hope to accomplish

- ◯ Yes
- ◯ No
- ◯ Other (please specify) - If possible, keep option open to return

Cost of Water

How much do you pay for water per month for your household?

____

Don't Know

Not Applicable

Water Use

How is this water used?

Drinking water

Cooking

Handwashing

Washing clothes

Agriculture - Include small garden plot or "kitchen garden"

Livestock

Other (please specify) ______________________________

Don't Know

Not Applicable

If you drink this water, do you treat it before drinking?

◯ Yes

◯ No

Not Applicable

If this water is treated for drinking, what method(s) is used?

Sawyer filter

Chlorine tablets

Chlorine liquid

Aquatab

Boil water

Sun exposure (UV)

Other (please specify)

Don't Know

Not Applicable

In this household, how many people use this water?

____

Don't Know

Not Applicable

Are any of the household children younger than 60 months old?

◯ Yes

◯ No

If "Are any of the household children younger than 60 months old?" is Yes, ask

How many children younger than 60 months of age (less than 5-years-old) are there in the household? ________________________________

If "Are any of the household children younger than 60 months old?" is Yes, ask

How many of the children under 5 years old had loose stools 3 or more times a day in the last two weeks? ________________________________

Not Applicable

Administer Rapid Diagnostic Tests and then respond: How many of the household members have malaria?

____

If Are any of the household children younger than 60 months old? is Yes, ask

How many of the children tested positive for malaria (put "0" if none were)?

Not Applicable

How much did everyone in the household spend on diarrhea treatment in the last month?

____

How much all together did everyone in the household earn in the last week?

Water Point Reliability

Was there a time in the last YEAR when you did not have sufficient water from this water point?

- ◯ Yes, at least once that was functionality related - A part or piece broke or was in disrepair. If this is selected, please explain what was broken and for how long below
- ◯ Yes, at least once that was weather related - If this is selected, please explain when below
- ◯ Yes, at least once that was related to something else - If this is selected, please explain what happened below
- ◯ Yes, it is currently not providing sufficient water - If this is selected, please explain why below
- ◯ No, always sufficient water
- Don't Know
- Not Applicable

If water is collected from off-site, who collects the water?

- Women collect the water
- Girls collect the water
- Men collect the water
- Boys collect the water
- Staff collect the water
- Paid workers collect the water
- Other (please specify)
- Don't Know
- Not Applicable

How long is spent collecting water for the house in a day, usually? Hint: Include round-trip travel = queuing time

- ◯ Less than 15 minutes
- ◯ 15-30 minutes
- ◯ 30-45 minutes
- ◯ 45 minutes to 1 hour
- ◯ 1-2 hours
- ◯ Other (please specify)

Don't Know

Not Applicable

Water Point Functionality

If the water point is on site, is water available on the day of the survey?

- ◯ Functioning
- ◯ Partially functioning
- ◯ Not functioning
- ◯ No longer exists

Don't Know

Not Applicable

Sanitary Inspection

Is this water point protected from unwanted intrusion by animals?

- ◯ Yes - Usually a fence perimeter or inside a structure
- ◯ No
- ◯ Other (please specify)

Don't Know

Not Applicable

Is there a cement plate around the pump that drains away from it?

- ○ Yes - Do not want muddy ground below water point
- ○ No - Puddled water or muddy under water point
- ○ Other (please specify)

Don't Know

Not Applicable

Water Quality

What color is the water?

- ○ Clear
- ○ Yellow or tan
- ○ Green
- ○ Brown
- ○ Other (please specify)

Don't Know

Not Applicable

Is there any visible material floating on the water or settling to the bottom?

- ○ Yes, material floating to the top of the water
- ○ Yes, material settling to the bottom of the water
- ○ No visible material in the water
- ○ Other (please specify)

Don't Know

Not Applicable

Is the water different colors during the year?

- ○ Yes
- ○ No

○ Other (please specify)

Don't Know

Not Applicable

Does the water have an unpleasant odor?

○ Yes

○ No

Don't Know

Does the water have an unpleasant taste?

○ Yes

○ No

○ Other (please specify)

Don't Know

Has the water been tested for bacteria?

○ No, not tested

○ Yes, tested negative - Bacteria not found

○ Yes, tested positive - Bacteria found

○ Other (please specify)

Don't Know

Sanitation

Is there a toilet or improved latrine on site?

○ Yes

○ No

Don't Know

If no, how far away is the nearest toilet or improved latrine from the water point?

○ 10-20 meters

○ 20-50 meters

○ More than 50 meters

○ No toilet or improved latrine nearby - i.e., Walk home, or go in the bush or field

○ Other (please specify)

Don't Know

Not Applicable

Type of Latrine

○ VIP Latrine

○ Flush/pour

○ Pit latrine

○ Composting toilet

○ Bucket

○ Hanging toilet

○ Other (please specify)

Don't Know

Not Applicable

If "Are any of the household children younger than 60 months old?" is Yes:

Where do children defecate in your household? How are the child's feces disposed?

Hint: Do not read all the options to the respondent, instead ask the question then choose the closest options below. Use the not applicable option if they do not have children.

Child defecates in the toilet

Child defecates in potties/nappies and feces are disposed in the toilet

Child defecates in potties/nappies and feces disposed anywhere

Child defecates in an open place and feces is picked up and disposed in the toilet

Child defecates in an open place and feces picked up and disposed in a dumping site

Child defecates in an open place and feces is covered with soil

Child defecates in an open place and feces is disposed anywhere, or left where it is

Not Applicable

Hygiene

○ Tippy Tap with soap

○ Tippy Tap without soap

○ Tippy Tap with ash but no soap

○ Yes, fixed facility with soap - sink or tap

○ Yes, fixed facility with alcohol-based rub - sink or tap

○ Yes, fixed facility but no soap or alcohol-based rub - sink or tap

○ Yes, mobile facility with soap - bucket, jug, kettle, tippy tap

○ Yes, mobile facility with alcohol-based rub - bucket, jug, kettle, tippy tap

○ Yes, mobile facility but no soap or alcohol-based rub - bucket, jug, kettle, tippy tap

○ No handwashing facility

○ No permission to see

○ Other (please specify)

Don't Know

Not Applicable

Waterborne Disease

Is there waterborne disease in this area? (If yes, please explain in the comments section which diseases are known)

Yes in our community

Yes in our family

Other (please specify)

Don't Know

Not Applicable

WASH Program

Is there a WASH program in your community such as a WASH Committee, WASH Management Committee, or Health Advisory Committee?

◯ Yes

◯ No

◯ Other (please specify)

Don't Know

Not Applicable

What is the best way to contact you or another family member that would like to become involved in a WASH Program, if applicable?

____

Hint: Name and contact

Survey phase

The survey phase is used in the analysis and visualization of results to compare the conditions before, during, and after project completion. If there is no current project taking place, choose Monitoring. Refer to the following definitions for more information:

Baseline - typically performed immediately before the beginning of a project, a baseline survey is what future surveys will be compared against to measure the changes that have occurred in the community; it can also help to guide the project planning and implementation.

Midline - usually performed during the middle of a longer project, it can help determine if trends in indicators are consistent with the expected outcomes of the project.

Endline - performed immediately after the end or handover of the project. This is used to determine if the outcomes were achieved.

Monitoring - may be performed at any time during or after the project. It can be used as a management tool that leads to learning and problem resolution, or to measure the long-term sustainability of the change brought about by the project.

In an impact evaluation, it is common to also perform the baseline and endline surveys in nearby communities that are not affected by the project. This data can be used to determine what changes would have occurred anyway in this region without the project.

- ○ Baseline
- ○ Midline
- ○ Endline
- ○ Monitoring

That's all! Thank the person for speaking with you and say Goodbye.

Chapter 8: Contract with the Community and Partners

We have worked a lot on developing a multiagency and community contract that works.

It is important to have contracts or written memoranda of understanding that specify purposes and expectations from each partner and timeframes.

What we have developed to date follows:

Agreement among Community, local NGOs and Rotary Clubs, and international partners

The purpose of this effort is to improve life in (village name). The effort is led by the community with help from local and international partners.

Our goals are to eliminate diarrhea, malaria, respiratory infections, and birthing deaths in (village name) by doing the following:

> Make water available closer to households, through wells with handpumps—women and girls will have to walk less to obtain and carry water, saving time;
> Water quality will be improved, helping to stop diarrhea, and lessening expense in treating diarrhea;
> Malaria will be prevented by malaria nets, diagnosed by rapid diagnostic tests, and treated immediately in the village, decreasing the spread of malaria; and
> Income will be increased through support of village savings and loan associations (VSLAs) and saving of money due to treatment of malaria, colds, and diarrhea.

Proposed community contributions to the projects in (village name) ____________:

WATER

> Provide food for the engineers and drillers
>
> provide accommodation for the engineers and drillers
>
> provide security
>
> provide sites for the wells
>
> provide materials e.g., bricks, river sand, and stones

provide repair and maintenance

LATRINE

Dig the pits and cover it

Install tippy taps

Provide sites for bricklaying

Provide doors, walls, and rooves for the latrines

MALARIA

Provide storage

provide at least 2 volunteers to help in the treatment of malaria and Rapid Diagnostic Tests (RDT) tests.

PROCESS

Conduct baseline survey.

Plan with community and local partners.

Contract construction where appropriate.

Implement, latrines and Tippy Taps first, P2P Inc and CCEDUC supplying design, cement, reinforcing iron, and 4" ventilation pipe once pits are designed.

Villagers build their latrines with the training and support of our liaison.

Once the latrines and tippy taps are in place, we do wells fast.

Once village health teams (VHTs) are in place, we buy anti-malarials and RDTs and nets the health office cannot supply.

Plan for sustainability.

Conduct final survey.

Discuss and evaluate benefits with communities and partners.

Local partners will provide leadership in planning, supervision, and evaluation of the project.

International partners will support the community's and local partners' efforts with additional funding, technical support, and guidance, as necessary.

Signed and dated: Community leadership, women leaders, and heads of families, local and international partner representatives sign and date the agreement.

Chapter 9. Latrines and hygiene, 100% coverage

The reason for sanitation is that many diseases are transmitted by fresh feces, especially species-specific feces. In general, the diseases are only transmitted by other people's feces, not one's own. So, the objective is to prevent contact with the fresh excreta if other people and to achieve 100% coverage in communities. Leaving anyone without sanitation means they will be the ones who can transmit diseases.

Sanitation can be accomplished by burying feces, flushing it into sanitary systems or septic tanks, composting it, or otherwise killing the organisms in the excrement.

Each of the many options have different costs, generally resulting from their complexity and the amount of work and infrastructure required to construct the systems.

Generally, since disease transmission is from other people's feces, reiterating, we strive for 100% coverage. It is not enough that one build a latrine for oneself or one's family. The entire community and those upstream need to have sanitary disposal systems. That is because these diseases are environmental, not individual. The disease is transmitted by the conditions in which people live. Fix the conditions and the diseases go away.

Some of the many diseases that are transmitted by feces include many varieties of diarrhea, worms, and other parasites. Polio, cholera, typhoid, Ebola, trachoma, guinea worm, schistosomiasis, ascaris, hookworm, and shigella are among the many fecally transmitted diseases. The US Centers for Disease Control list about 90 diseases transmitted by feces.

There are many considerations in construction of sanitary facilities. How do people customarily defecate—squatting or sitting? How do they clean themselves? Do the facilities produce odor? Odor is associated with disease transmission in some cultures. Do they produce flies? Flies can carry enough of some bacteria to cause diarrhea. Do they have hand-washing access? Do they have soap or sanitizing services? Additionally, space, cost, and the water table are considerations.

Some of the sanitary options follow:

Pit latrines;

Ventilated improved pit latrines;
Pour-flush latrines;
Pour-flush to sewer systems;
Ecosan latrines;
Composting latrines;
Tiger worm composting latrines;
Septic tanks; and
Sewer systems.

Each of these options entails different costs and construction competencies. A pit latrine can be constructed for a few dollars. Each of the other models costs more, ranging up to thousands of dollars to build.

Planners need to consider all these factors including national standards in selecting sanitation models.

In this book, we will share how to make the ventilated, improved pit (VIP) latrine. The reasons are its low cost, high acceptability, and effectiveness in preventing bacterial and worm transmission. It does not work where there is a high water table or multi-story dwelling with no yard space.

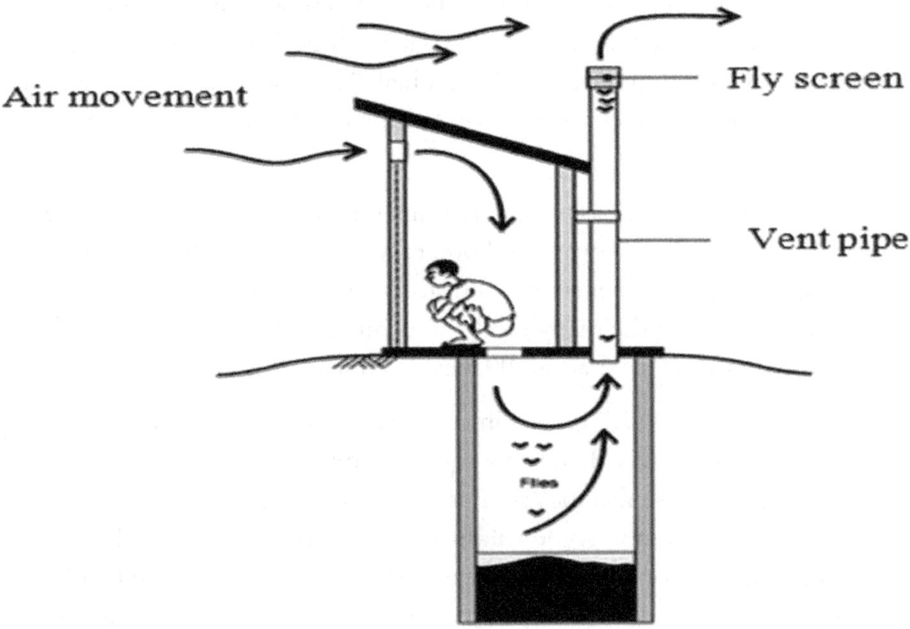

VIP design

To build a latrine, locate it at least 15 meters from water sources. Follow the national standards. Otherwise, it is desirable to have the latrine close to one's dwelling. Various pit designs can be used. They influence the cover plate dimensions. Typically, the pit is 2 meters long by one meter wide and at least 2 meters deep and less than the water table at its highest. If the walls tend to cave in, the pit needs to be lined by bricks or other support measures.

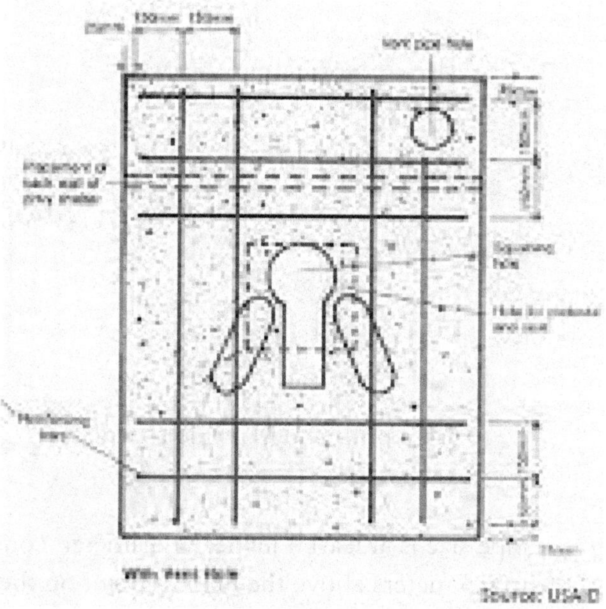

Baseplate

The amounts of materials required to make this cover plate, according to the World Bank, follow:

Cement 25 kg.,
river sand 0.125 m3, and
1.5 kgs of 3 mm. reinforcing iron.

Mass produce latrine cover plates when possible

Mass producing Latrine baseplates

The best vent pipe size is at least 4 inches in diameter, connects directly from the hole to 0.5 meters above the highest point on the roof, is covered by a fly screen, and allows direct sight of the sky from the pit (it has no cover against rain nor elbows).

Due to the Venturi effect, the pipe sucks the odor out of the cabin, through the pit, and out the end of the pipe. The result is an odorless pit.

Furthermore, flies in the pit see sky through the vent pipe but a roof and obscure cabin through the defecation hole. Flies fly to the light, up the pipe, not into the cabin. Tests showed no contaminated flies in the vicinity of latrine pits with this model.

Roofing, door, and wall materials can be varied to use the least expensive. The defecation hole requires no cover: air circulation is desired.

Cover plate and walls layout of VIP latrine

Children wash their hands using Tippy Taps in front of large family latrine

The other element of stopping feces transmission is washing hands with soap after using the latrine. A frequent problem is the person with contaminated hands touches the container or the faucet to turn it on, spreading bacteria to it, washes his or her hands, then touches the contaminated lever or container to turn off the water, resulting in re-

contamination. For these reasons, we work on setting up touchless handwashing facilities with villagers.

Many touchless solutions including the Tippy Tap (shown in previous image) have been summarized by UNICEF. Construction and purchasing guides can be found on the internet.

The additional factors that need to be addressed are drainage, so the water does not pool on the ground and breed mosquitoes, and soap or detergent for washing. Studies have shown that soap needs to be used to remove bacteria.

Drainage for handwashing systems can be provided by digging a pit below the faucet, filling it with gravel, and covering the top with sand. The water quickly goes underground.

Chapter 10. Monitoring, Evaluation, Research, Learning and Continual Improvement

The key purpose of monitoring is to optimize your work. Objectives are to keep field staff, suppliers, partners, and the community engaged; to solve problems that occur; to continually improve the process; and to test alternatives.

Remember that when you plan a community-change process with communities and partners is the time you know least about the situation. Once you start buying, shipping, storing, delivering, doing the fieldwork, and working in each community, you will find out many things you did not know before. Monitoring partially lets you know what the reality in the field is and allows adjusting to it. It also facilitates the increase of motivation and commitment.

When doing field-monitoring, use some of the following:

Ask to whom you are speaking, their name, function, etc.;
Introduce yourself and where you are from and why you are there, briefly, and humbly;
Ask what successes they are having in their work and what innovations they have developed—congratulate them;
Ask what problems they are encountering: can they solve them and what would the solution look like?;
When solutions are not in the control of local implementors, adjust the project as appropriate to solve problems; and
Make plans and commitments to solve the problems: note them, set dates and follow-up.

Good sanitation and hygiene is a first step to solving fecally transmitted diseases. Next, we address water.

Chapter 11. Safe water and maintenance plan

Many options are available for providing safe water. By "safe," we mean water that does no harm to the user, not transmitting disease, illnesses, nor poisons. The options vary by environment, population density, and locally available resources, materials, skills, and equipment.

For any community to exist, it already has some type of water supply. Drinking water can be chemically de-toxified or filtered to remove bacteria. Not all filters are equal, though. Filtering must be with smaller pores than the size of bacteria.

The community water may be inadequate in quality, quantity, or accessibility. WASH projects should address factors including proximity, affordability, sustainability, quality, and quantity as cost-effectively as possible. A checklist for designing and reviewing community WASH projects is in the Appendices on page 82.

Some of the many options in providing water follow:

Spring or surface water captation and gravity flow systems;
Rainwater collection systems;
Wells, both boreholes and hand-dug;
Dams and ponds;
Desalinization systems;
Water reuse systems;
Atmospheric condensation;
Community or household water treatment systems, and
Storage and pumping systems.

Wells:

There are hand-dug wells and boreholes. Their prices vary five-fold. Hand-dug wells frequently run dry in the dry season, especially if they are dug during the rainy season when the water table is high. Consequently, dig at the end of the dry season when the water table is lowest.

Wells and pumping systems need to be protected from contamination. They need to be secure against groundwater flowing into them, contaminated buckets or lift systems, and diapers being disposed in them (that was the cause of cholera in London that Dr. John Snow found in 1854, establishing the discipline of epidemiology).

Pumps are mechanical so subject to breaking. Some, like the rope pump, can be maintained with materials locally available. Others, like solar-powered submersible pumps, cost more and may require specialized skills and national or international parts. A friend told me of one that was struck by lightning, shutting down the expensive system. Some lever-operated well pumps have been tested for decades, are relatively reliable, and have locally made parts. All these factors need to be considered when choosing pumps.

Additionally, when using subterranean water supplies, they need to be tested for quantity and quality. Simple coliform tests indicate fecal bacteria in water. Mineral tests are more expensive. Heavy metals can cause diseases, cancer, and death. One of the major water poisoning episodes occurred when thousands of boreholes in Bangladesh poisoned over 40 million Bangladeshis with arsenic.

Filters: there are several highly effective point-of-use household filters, including

> Sawyer's and many other companies' nanotubule filters;
> Biosand, filters; and
> Silver nanoparticles pot filters.

These point-of-use household filters achieve over 99% removal of bacteria, usually enough to stop disease transmission. Some studies indicate they may eliminate diarrheal incidence, although that seems illogical. It does indicate an impressive effectiveness. Some of the challenges are the weight and equipment necessary to make biosand filters, the fragility and two-year useful lifespan of the pot filters and other variations, and the maintenance training required and foreign manufacture of the nanotubule filters. On the positive side, they all can greatly reduce the incidence of diarrheal and intestinal parasite diseases for as little as $10 a household, according to how the program is designed.

The process to develop a water supply system for a community suffering from diarrhea and inadequate water supply is relatively simple. Consult the local water officer. Look at local water systems and examine their pros and cons. Check technologies that are available in the area. Select a sustainable, affordable, adequate system. Get three quotes from experienced, successful local suppliers. Sign locally recommended contracts, paying only part of the money down. Ensure the contract includes conditions for testing quantity and quality and conditionality for completion. Reserve part of the payment for satisfactory completion.

A father in rural Uganda gives his daughter Sawyer-filtered water:

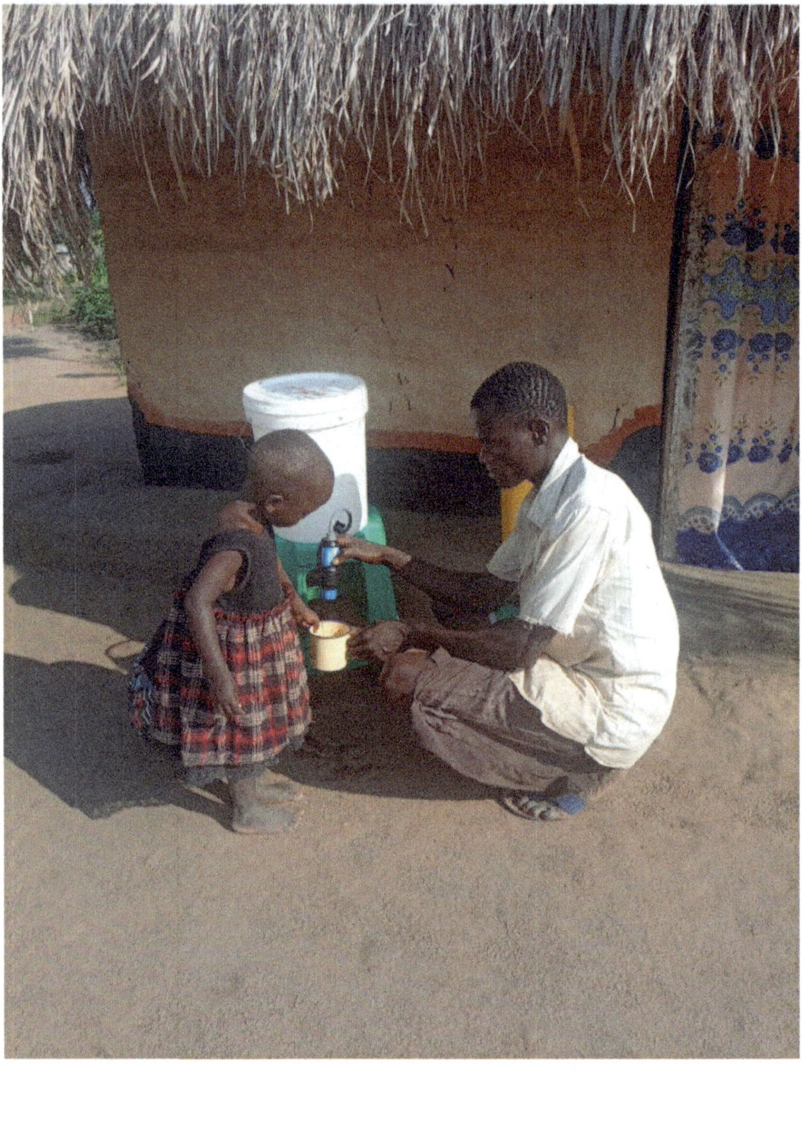

Completed deep borehole in rural Ugandan community

Chapter 12. Malaria elimination with LLITNs, RDTs, 100% treatment, looking at Mass Drug Administration

Malaria is another water-related disease. It is disastrous for hundreds of millions of people. It results in deaths, serious illness, anemia, medical and drug expenses, hospitalization, and lost workdays. As much as 20% of the household income is commonly spent in treating malaria.

Malaria is preventable and curable. It used to exist in North America, Europe, Japan, and Korea, but we have eliminated it. Malaria can be eliminated.

To get rid of malaria, we need to understand its host and transmission. Commonly, the only organism that the parasite (plasmodium) can multiply in is the human. If no humans have it, malaria ceases to exist.

The Anopheles mosquito transmits malaria from infected individuals to others by sucking blood. The mosquito breeds in swampy, stagnant water. It is a night-time mosquito. It can fly about 5 kilometers when filled with blood, but winds can carry it greater distances.

Therefore, eradication of malaria can be done by eliminating the mosquito's breeding grounds or preventing its access to infected or well people. These things are accomplished by preventing water pools from forming, draining swampy areas, introducing fish to pools of water, using mosquito nets, especially Long-Lasting Insecticide-treated Nets, spraying home interiors and around homes with insecticide, wearing long sleeves, hats, long pants, socks and shoes after dark, and using mosquito repellents. One hundred percent DEET repels mosquitoes.

Additionally, fully treating infected people as soon as possible after infection decreases the spread of malaria. For that, rapid diagnostic tests (RDTs), full doses of effective treatments (commonly Artemisinin Combination Therapy or ACT) being available, and people taking full doses.

Work with national experts to develop eradication plans. Possibly the best activity of non-governmental agencies is piloting alternative approaches to find ways to improve national programs.

To develop a malaria eradication program, follow these steps:

Check with the local health office and the community what the situation is;

Develop a pre- and post-household survey that addresses demographics, mosquito nets, malaria incidence, and practices (contact for drafts I have used);

Develop a control and eradication plan and consult local authorities and the communit(ies) to improve it;

Develop a local and donor initial and sustainable budget;

Sign agreements;

Train, survey, and implement;

Survey impact;

Evaluate with all stakeholders, drawing conclusions of lessons learned and future strategy; and

Develop a sustainability plan.

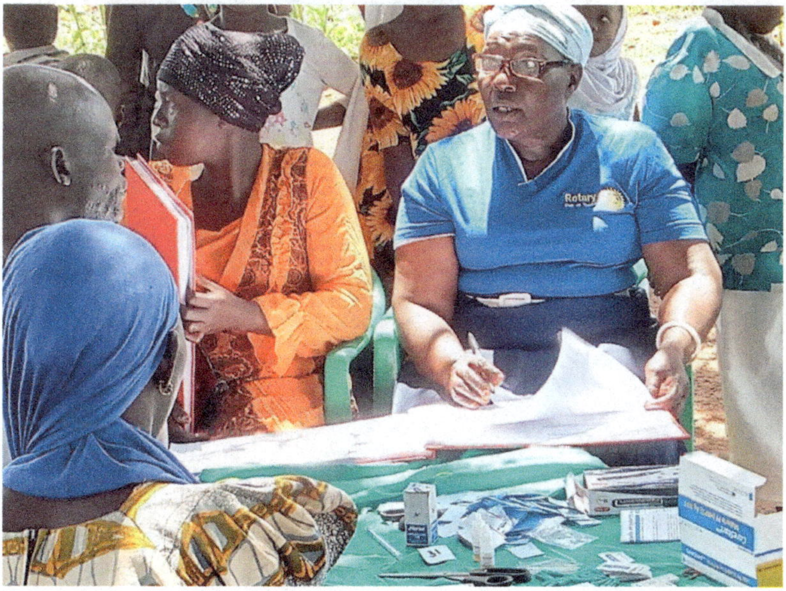

Yumbe Rotarian Lenny Christine conducting Rapid Tests then distributing anti-malarials to villagers

Yumbe Rotarian Lenny Christine giving malaria talk to villagers

Chapter 13. Microloans and sustainability

Two extremely popular and impactful programs are savings and loan groups (Village Savings and Loan Associations, VSLAs, vsla.net,) and microloan groups.

The VSLA Field Officer manual gives instructions how to form a village savings and loan association for a group, typically 30 local entrepreneurs. Members contribute money monthly to a savings pool. After a few months, a few members borrow the money, conduct a profitable enterprise, and pay the money back at an agreed interest and payment rate. In Muslim countries, the interest rate is called an administrative fee.

Listening to stories and analyses by these groups, we have learned they use their pooled resources both to make money and as a social security fund, money available for hospitalizations, school fees, and funerals. Savings are small, reassuring, and opportunities for developing group trust and solidarity.

If the group carries on the VSLA activities successfully for a year, it is good to consider upgrading it into a microloan association. Given additional funds to administer and loan out or form a VSLA cooperative, members can make far more money, upgrade their living conditions, educate their children better, and prevent sporadic hunger. The pool of money grows relatively rapidly, and members improve their living conditions impressively.

One microloan program, TCP Global, that was started in about 2000 has been particularly successful. It has formed a Rotary Club, is starting a cooperative program, typically doubles incomes, and stops episodic hunger, according to surveys, reporting, and data.

TCP Global signs agreements with local NGOs that can receive international fund transfers. The NGOs identify local successful VSLA groups, sign agreements with them, then pass money transfers from TCP Global to the VSLA groups.

The groups administer the funds, maintain records, and report monthly or less frequently if money is lent or paid less frequently. The method is to keep all money lent out. When money is paid back, it is immediately lent to a new borrower, avoiding lock boxes, banks, and unused funds.

Key operating practices of TCP Global follow:

To decide how many members you want in your group, first select from people who are already running small businesses but would like to grow. Second, identify the average loan amount people want. Divide the money pool by that amount and multiply by 2. That is the group size that will be most successful for you.
Next, there are seven major TCP Global rules:
To start the loan process, request an agreement contract, sign and date it, and return it;
For microloans to members, charge the current local bank interest rate for businesses;
Keep the money loaned out: when it comes in, try to loan it out at the same meeting rather than holding it in a lockbox or bank account;
Try to ensure everyone repays her loan on schedule: no matter what, make sure 95% repay;
Whenever there is a transaction, keep a record and, as soon as possible, record it on the Daily Log spreadsheet, only one transaction per line. Include the interest or administrative fee amount in the payment. Some instructions about the spreadsheet follow:
To understand this information, open a Daily Log spreadsheet in Excel.
On the bottom, click the "Example" tab;
In cell R1, set the language to your preferred language.
Notice that the row 2 has "Funds Received from US" in column B, a date in column C, and amounts in column D and E filled in. D2 is the amount deposited in your bank account in local currency. E2 is always $1500. Everything else fills in automatically.
Row 3 starts the community loan process. Write one person's name, the date, and the amount she received in columns B, C, and F. Everything else is filled in automatically by the spreadsheet. Entries in column F are automatically subtracted from the balance on hand.
Repayment records start after all the money has been loaned out. In the case of this spreadsheet, they start with row F. Entries in column G are automatically entered back into the balance.

Cells Q2 through R11 automatically calculate summary statistics for your spreadsheet. h. Try to keep all the money loaned out, so that the Balance in R4 is 0 or close to it.
Whenever the Loans in R9 equal twice the amount received in R7, the program can request an additional sending of money from the USA. Ask a TCP Global volunteer if anything else is unclear.
When there have been transactions on the spreadsheet, send the spreadsheet to TCP Global at least once a month; and
When the group has made microloans of twice the amount received from TCP Global, if you want more money, request an additional transfer from TCP Global. Not doing any one of these things may result in TCP Global stopping to work with your group.

Following these simple rules, communities, especially groups of local women entrepreneurs, can quickly improve their lives and well-being.

More than 30 microloan groups of 30 or more women members each celebrated the arrival of the TCP Global executive director, Helene Dudley, in Yumbe, Uganda

TCP Global supports this process with pre-designed reporting spreadsheets, automated analysis systems, and internet-based record systems. TCP Global works with Rotary clubs.

Examples of business activities by women include beauty parlors, market

pharmacies, selling of produce, farming, animal breeding, and many others. A few photos follow:

Chandiru Emily sells drugs

She uses her profits to buy and raise ducks

Entire groups of women benefit from microloans to their members

Women's microloan member in Niger profits from work as a seamstress

Chapter 14. Final survey for impact

Basically, the final survey can be the same as the initial one. It needs to include the date and place. The comparisons can be two-fold: one, compare the initial to the final situation, measuring how great the changes have been; the other is comparing similar communities' situation at a point in time. The first is a pre- post- evaluation. The second is a case-control evaluation. The latter is considered more accurate in that other factors than the proposal may have produced the changes.

In the evaluations, one can decrease work by drawing a sample. The best samples are random but systematic sampling can be done. The required sample size depends on how large a change is reasonable to expect and how large the population is.

Our goals in water-related projects are elimination of diarrhea, malaria, hunger, excessive labor, absenteeism, and fewer expenses. We measure goals we plan to achieve, designing before and after surveys to track our effects. The summary of a recent survey that was done of a community we worked in follows:

Changes resulting from Achiba, Uganda WASH and malaria Program w/RPCVs, iNGOs, Yumbe Rotarians, finished in August '22, compared to August '21:

> 90% less diarrhea: 50% to 5% of children had diarrhea in last two weeks
> 52% less malaria: 75% to 36% of everyone
> 90% less money spent treating malaria and diarrhea: from 31% of household income to 3%: over $20,000 of savings from being disease free per year.
> Water no longer dries up during dry season: 80% to none
> Total latrine coverage: 35% to 100%
> Total hygiene coverage: 10% to 100%
> Community collects water and malaria program maintenance fees
> Mobilized VHTs and Village Water Committee
> Rotary Community Corps was formed and is active.

Chapter 15. Other Worthwhile Components to Consider Adding

Makiga brick- Makiga Machine used for making bricks for the construction of latrines without destroying the environment:

Woman making soil stabilized interlocking bricks with Makiga machine

Interlocking soil bricks drying before use

The benefits of the soil-stabilized interlocking bricks are many. Not

requiring firing, they decrease the use of firewood and cutting down trees. Being uniform in size and interlocking, they decrease the quantity of mortar used. They decrease expenses in bricks, mortar, and environmental degradation.

Reusable sanitary pads and diapers

In elementary school, girls frequently are withdrawn to carry water for the family, herd animals to and from water sources, or take care of sick siblings. Building better water supplies closer and decreasing diarrhea helps keep younger girls in school. In high school, girls are taken out of school because of the embarrassment of menstruation without pads. Making reusable sanitary pads increases girls' chances of staying in school and decreases wastes. Good Foundations International, a partner, trains local women to make both reusable sanitary pads and diapers.

Women making sanitary pads from GoodFoundationsInternational.org pattern, instructed by executive director, Jennifer Mally

Making sanitary pads by hand

Exclusive breastfeeding until six months of age:

Exclusive breastfeeding for the first six months of life (no water, juices, or foods other than breastmilk) starts newborns to excellent weight gain and health. It prevents diarrhea, protects the infants from diseases, and bonds the babies and mothers.

Oral rehydration therapy when diarrhea occurs:

Diarrhea does not kill. Dehydration kills. The World Health Organization, UNICEF, and national expert organizations recommend drinking as much liquid as is lost during diarrhea and vomiting episodes to replace lost liquid. They recommend if possible the use of special formulations of sugars and salts or complex carbohydrates and salts. Never add more than the indicated amounts to the volume of water that is specified. Continue to feed during diarrhea and vomiting.

School activities and programs:

Schoolchildren love to dance, sing, recite, and act for their families. Parents and neighbors attend their performances in large numbers. An easy way to show the relevance of WASH and disease-prevention activities is to have the children sing, dance, recite, perform, and even report on them. Providing teachers with key messages for these performances can allow easy child and community education and promotion of the programs.

School WASH services can be upgraded as part of good community WASH programs. Important considerations are adequate water supplies, latrines, handwashing facilities with soap, menstrual hygiene, and solid waste disposal. See appendix on school WASH criteria, page 89.

Environment:

The environment is suffering due to global warming and deforestation. Coupling tree planting, using government nurseries to obtain seedlings, and planting, watering, and protecting trees at households is an easy add-on activity to water programs. Additionally, the water can be used to grow mosquito-repellent and treatment plants and high nutrient-density vegetable gardens, especially of dark green leafy vegetables, near homes. Preferably, choose what is most acceptable to the population and what they know how to prepare.

Maternal child health:

The people who suffer and die most from water-related diseases are young children. Planning the program to prevent their infection and ensure their rehydration decreases suffering, costs, lives lost, and money spent by over 70%. Additionally, the WASH services can be extended to health centers and hospitals, further decreasing disease transmission and iatrogenic infections (see appendix on health and hospital WASH criteria, page 85). Additionally, producing and consuming dark green leafy vegetables daily overcomes many problems.

Community phone and solar charger for birthing, emergency calls, and coordination:

For program communication and coordination, it is useful to ensure communities have smart phones and solar rechargers when they lack electricity. At the same time, communication helps birthing mothers and injured or dying people to reach ambulance and transportation services, decreasing birthing losses and other causes of death. Additionally, the phones can be leased for commercial uses, allowing villagers to find what different markets pay for their produce and what different suppliers charge for inputs. This increases incomes.

Ajaga Innocent Buran, executive director of CCEDUC in Yumbe, shows smart phone and solar panels purchased for $108 to coordinate with a village

The Makiga brick machines, reusable sanitary pads and diapers, tree, herb, and vegetable planting, and phones are useful add-ins to water programs if their addition is manageable.

Chapter 16. Concluding Whole Systems Change Conference

The final conference is done the same way the initial one was, but its objective is for the stakeholders to conclude what they have learned, what changes they are committed to making occur in the system, and if they have next steps in development they plan to undertake.

According to W. E. Deming's theory, any problem that exists in more than 2 standard deviations from the mean (about 3% of the population) results from the way prevailing systems are set up. To overcome the problem, stop working on individual cases and fix the system. This approach resulted in the revolution in manufacturing that started in Japan after WW II and became total quality management, Six Sigma, and Xerox's stakeholder team problem solving process (). Implementing systems change is done through continual efforts, data collection, evaluation, and modification called action research ().

The goal is to find and demonstrate better ways to set up societal and governmental systems. These approaches inform our efforts in WASH, malaria, hunger, and poverty. We involve those who suffer from deprivation and other key stakeholders who control the system, consult them, analyze the system, come up with solutions, and document the solutions.

In this way, rather than continually doing projects, we strive to change the systems that produce disease and poverty.

Chapter 17. Conclusions

Good health begins in the community, environment, facilities, and practices. It is important to accomplish these improvements with communities and partners in an impactful, sustainable manner. How we do the work is as important as the technologies we use. In this book, we have attempted to share the best practices and resources we know to build impactful, sustainable WASH systems.

Summarizing, villages want good water service. They are willing to organize and work to get them. Once in place, if designed appropriately, the systems can result in several times their investment cost in the first year but can function for ten years or more. Benefits are decreased disease, death, and labor, lower absenteeism, and greater production and income. We are more likely to get those benefits if we design for them and measure them.

For lasting progress to occur, it is better that the communities ask us to work with them, not that we find them and offer help. Development is about empowerment, not dependency generation. The communities need to organize, request, commit, and manage their improvements. Partner with local organizations and communities ready to do that and use each experience as an opportunity for them and us to learn better processes and technologies.

Partnering with local non-profits, Rotary Clubs, and governments should decrease costs and increase learning and efficiency. It is also critical to a major goal, systems change.

Amazing new possibilities have been opened with cell phones, the internet, and group calls. Nowadays, we can plan, organize, coordinate, survey, monitor, and learn using tablets, cell phones, and the internet. Meeting with partners weekly, listening to reports on action plans, and continually learning are terrific possibilities. To conceptualize their conditions and situations, to bond, it is helpful to visit in person, but, once good partners are found, it is feasible to partner internationally using the abundant and growing resources of the internet. Our weekly Zoom meetings and WhatsApp calls keep me informed and engaged.

It is important to budget when we plan, committing partners to pay for different parts of the work. We shared one of our budgets, but budgets can vary widely. Some organizations do sanitation with no external funds, but I have never heard of one that gets 100% coverage, the requirement to prevent fecal spread of disease. The budget depends on local costs and materials, what is available in communities and their incomes, and the technologies chosen. For instance, borehole rehabilitation typically costs a quarter of new drilling.

After a community applies for help, we have local Rotary Clubs or other partners follow-up with them to ensure the needs were correctly reported and that the communities are motivated to work with us. This requires an in-person visit, usually by a small team.

The next step is a community survey of the village. The survey can be a sample or a 100% household survey. In developing or selecting surveys, it is important to use standardized indicators when available and measure what we plan to change. Surveys need to be field-tested, finalized, checked for data quality, and used both with partners and the villagers. Measuring ensures the quality of the work and increases learning.

Before starting work, all organizations and the communities should sign a memorandum of understanding stipulating what each will do. It is important to be clear on commitments and timeframes, using a who, what, where, when, and how framework. Sequencing of the work needs to be clear, too. We like every head of household to sign the document along with our partners.

In the sequencing, we have found it is best to start with sanitation and hygiene, requiring 100% coverage before beginning water. Starting with water has resulted in non-compliance with the rest of the project, in our experience. After sanitation and hygiene, we reinforce a water committee, the planning of operation and maintenance, training, and setting a water fee and payment schedule. Then we work with the community to put in an affordable, appropriate water system, signing a contract with the implementor stipulating all requirements and the payment schedule. Testing water quantity and quality is important. Provisions for what happens if water is not found should be included in contracts. Payments should be for costs incurred until the system is completed. Household filters or treatment can be added after the system is completed. For filters, operation, maintenance, and rehabilitation training are all important.

Another partner, Water with Blessings, has an excellent model for this training. To finalize water interventions, we do malaria eradication, also, because it is a water-related disease. Trachoma, worms, and other problems should be addressed if they are locally prevalent.

Donors usually require monitoring and evaluation of community work. Wise organizations do the monitoring and evaluation for continual improvement, the donors, and for the communities and other partners. When problems are found, fix them. Don't just collect and report the information: use it to improve. Rotary has shifted to "MERL," monitoring, evaluation, research, and learning. Since we never know the most about an effort before beginning, MERL and continual improvement should always be part of implementation. We collect information to use it, and one of the ways we use information is to improve. When we find out something, we respond. Monitoring questions are best designed to recognize the work and effort of others, to increase motivation, and to find out if the process is working like we had envisioned and, if not, to fix it. MERL becomes MERLCI: MERL and continual improvement.

Our malaria intervention is like sanitation: we need to achieve 100% coverage. The reason is that we get malaria from others. With immunizations, Rapid Diagnostic Tests, traditional and combination therapies, mosquito nets, malaria repellent plants, and stopping mosquito breeding, we can eliminate malaria. At present, it results in high incidences in many areas, suffering, death, large expenses, school absenteeism, and lost labor. It is another water-related disease we can eradicate.

In this book, we have added additional activities that are important to consider. One is establishment of Village Savings and Loan Associations and microloan groups. With little investment, these groups can double incomes. If we sign contracts with them, they can be an additional source of funds for maintaining water systems. Rural communities love them, and they eliminate hunger and provide some social security. Other activities to be considered with water projects are tree-planting, vegetable gardens, herb gardens for treatments, reusable sanitary pads and diapers, soil stabilized brick machines, cell phones and rechargers, and many others. A water project should be the start of progress for communities.

To complete the work with the community, take two final steps. One, measure the benefits of what has been done. Conduct a final survey that examines the indicators collected at the start. Analyze the results. Then, convene all partners to evaluate how systems need to be changed to make these benefits the norm throughout the region, not just in the community. What needs to be changed? The reason to do this whole systems change analysis is that any problems that occur with higher incidence than about 3% (two standard deviations from the mean) are system flaws, not personal problems. Fix the system and the problems go away. That is often the goal of our work: fixing flawed systems.

The appendices contain several important documents: an example of a community contract; and design and review criteria for community, school, and health center WASH programs.

The intention of this book is to guide Rotary Clubs and other organizations through effective WASH and water-related methods. Let us know improvements you make. Thank you! Chris.roesel@gmail.com.

Appendices:

Example of a community contract

Design and review criteria for community WASH programs

Design and review criteria for school WASH programs

Design and review criteria for health center WASH programs

Model community agreement:

Agreement among Achiba, CCEDUC, and P2P Inc.

The purpose of this effort is to improve life in Achiba, Mechu, Kuru, Yumbe, Uganda. The effort is led by Achiba with help from the Yumbe Rotary Club, CCEDUC and P2P Inc.

Our goals are to eliminate diarrhea, malaria, respiratory infections, and birthing deaths in Achiba by doing the following:

> Make water available closer to households, through wells with handpumps—women and girls will have to walk less to obtain and carry water, saving time;
> Water quality will be improved, helping to stop diarrhea, and lessening expense in treating diarrhea;
> Malaria will be prevented by malaria nets, diagnosed by rapid diagnostic tests, and treated immediately in the village, decreasing the spread of malaria; and
> Income will be increased through support of VSLA groups, increased healthy days allowing more workdays, and saving of money from prevented treatment of malaria, colds, and diarrhea.

Proposed community contributions to the projects in Achiba:

WATER

Provide food for the Engineers

provide accommodation for the Engineers

provide security

provide sites for the wells

provide materials e.g., Bricks, river sand, and stones

provide repair and maintenance

LATRINE

Dig the pits and cover them

Install tippy taps

Provide sites for bricklaying

Provide doors, walls, and rooves for the latrines

MALARIA

Provide storage

provide at least 2 volunteers to help in the treatment of malaria and RDT tests.

PROCESS

Conduct baseline survey.

Plan with community and local partners.

Contract construction where appropriate.

Implement, latrines and Tippy Taps first, P2P Inc and CCEDUC supplying design, cement, reinforcing iron, and 4" ventilation pipe once pits are designed.

Villagers build their latrines with our training and support.

Once the latrines and tippy taps are in place, we build or drill wells fast.

Ensure point-of-use water quality.

Once VHTs are in place, we buy anti-malarials and RDTs and nets the health office cannot supply.

Plan for sustainability.

Conduct final survey.

Discuss and evaluate benefits with communities and partners.

CCEDUC and the Rotary Club will provide leadership in planning, supervision, and evaluation of the project.

P2P Inc. will support the Achiba, Rotary, and CCEDUC effort with additional funding, technical support, and guidance.

Signed and dated: Achiba leadership, women leaders, and heads of families, Rotary Club President, Ajaga Innocent, Director of CCEDUC, Chris Roesel, President, P2P Inc.

Implementation and design guidelines for community water, sanitation, and hygiene activities:

> Rating system: 1-10, the challenge is that we should have weighting for each.
> Baseline findings may be project elimination criteria (e.g., water has poisonous minerals in it).

Health Center/Hospital WASH Project Design and Review Criteria

Relevant excellence identification criteria: appropriate sharps and fluid waste disposal, water quantity, quality, accessibility, worms and insects controlled, safe feces disposal, clean hands, menstrual hygiene management, process, replicability, and sustainability. Review state of the art and national guidelines, considering these criteria.

Weight up to the maximum score depending on qualitative assessment of the factor.

Plan	Possible	Actual
Community assessment, participatory planning, and periodic review by key stakeholder groups	100	
Safe and enough water*		
No coliforms (or at least complies with country standards)	20	
Water complies with country standards for dangerous minerals	10	
Plan for schedule, supplies and personnel for water quality testing	10	
Water adequate for healthcare facility cleaning	10	
Provides at least 3 liters of drinking water per staff and patient per day, filtered and/or chlorinated if required	10	
Provides at least 20 liters of water per person for cleaning per day	10	
Design for adequate drainage (surface, waste, and toilet waste)	10	
Funding adequate for sustainability (fees / funds adequate for preventive maintenance and repairs)	50	
Tools, supplies, and materials adequate for common repairs	10	
WASH responsible person identified and held accountable	10	
WASH system maintainer initial and refresher training	10	
Regular training of staff on infection prevention using WASH	10	
Succession plan for personnel	10	

Menstrual hygiene plan and facilities	10	
Latrines/Toilets		
*Appropriate design for technical solution for local conditions (no ground or surface water contamination).	10	
Disability Accessible Latrines and handwashing facilities in center	10	
Distance from water source is adequate for pit latrines and septic tanks (30m)	10	
Convenient access to waiting and examination rooms	10	
Insects and other disease vectors controlled	10	
No odor	10	
Adequate anal cleansing and menstrual hygiene provisions	10	
Enough facilities for patients, patient support, and staff	10	
Tools, supplies, and materials adequate for common repairs	10	
Hygiene		
Touchless (e.g., Tippy Taps, foot pedals, elbow, or knee wand)	10	
Available and adequate soap and cleaning logistics system	10	
Convenient access to hand washing stations for staff, patients, and support people, including mobile stations if necessary	10	
Located within convenient access from latrines, waiting, and exam rooms	10	
1 shower or bath/40 inpatients and for labor/delivery (comply with national standards)	10	
Plan for schedule, supplies and personnel for hygiene cleaning	10	
Laundry and sterilization systems as appropriate	10	
Waste Disposal Plan (Comply with national standards)		
Safe protection of sharps	10	
Safe disposal of body fluid wastes (incineration?)	10	

Safe sharps and waste disposal system (either pick-up by hospital; lined, protected pit for sharps and waste disposal; or high temperature incineration if available)	20	
Storm and greywater drainage	10	
Hepatitis B vaccination of staff who work with sharps and biological wastes	10	
Other		
Signed agreement with clinic and partners detailing obligations/expectations (Defined roles for clinic and club)	10	
Baseline and final evaluation includes measures of all applicable items (100%=50, decreasing by steps)	50	
Continual improvement plan and process (monitoring plan)	10	
Adequate Personal Protection Equipment and logistics, storage, and training	10	
Drug and food management systems	10	
Lighting and back-up power lighting and ventilation for surgery, delivery, latrines, and evacuation	10	
Plan for crowding	10	
Total (620 possible)	620	

- Rating system: 1-10, the challenge is that we should have weighting for each.
- Baseline findings may be project elimination criteria (e.g., water has poisonous minerals in it).

Attendance: Chris Bloore, Marcos Calvert, K. Neil Van Dine, Chuck Fisher, Henk Holtslag, and Chris Roesel

Resolution: Make further suggestions as necessary.

School WASH Project Design and Review Criteria

Measure	Poss-ible	Ac-Tual
Plan:		
1. Conduct a participatory planning meeting with all stakeholders (see Methodology | Future Search for suggested process guide and Rotary's WASH in Schools guide https://my-cms.rotary.org/en/document/guide-wash-schools for stakeholders): clarify needs, design, plan maintenance, continual improvement, sustainability, and financing.	100	
Water		
1. Microbiological quality of drinking-water. Escherichia coli or thermotolerant coliform bacteria are not detectable in any 100- ml sample.	50	
2. Treatment of drinking-water. Drinking-water from unprotected sources is treated to ensure microbiological safety.	50	
3. Chemical and radiological quality of drinking-water. Water meets WHO Guidelines for drinking-water quality (WHO, 2004b) or national standards concerning chemical and radiological parameters.	50	
4. Acceptability of drinking-water. There are no tastes, odors or colors that would discourage consumption of the water.	10	
5. Water for other purposes. Water that is not of drinking-water quality is used only for cleaning, laundry and sanitation.	20	
6. Water available 24/7.	50	
7. Basic quantities of water required. Day schools 5 liters per person per day for all schoolchildren and staff Boarding schools 20 liters per person per day for all residential schoolchildren and staff Nonresidential schoolchildren and staff 5 liters per person per day	50	

8. Additional quantities of water required. The following should be added to the basic quantities as necessary. Figures given are for day schools. They should be doubled for boarding schools. Flushing toilets 10–20 liters per person per day for conventional flushing toilets Pour-flush toilets 1.5–3.0 liters per person per day Anal washing 1–2 liters per person per day	10	
9. A reliable water point, with soap or a suitable alternative, is available at all the critical points within the school, particularly in toilets and kitchens.	20	
10. A reliable drinking-water point is accessible for staff and schoolchildren, including those with disabilities, at all times.	20	
11. One shower is available for 20 users in boarding schools (users include schoolchildren and residential staff). Separate showers, or separate showering times, are designated for staff and schoolchildren, and separate showers or times are designated for boys and girls. At least one shower should be accessible for females with disabilities and one for males with disabilities.	10	
12. Laundry facilities, with soap or detergent and hot water or chlorine solution (or both), are provided in boarding schools.	10	
Hygiene promotion		
1. Correct use and maintenance of water and sanitation facilities is ensured through sustained hygiene promotion. Water and sanitation facilities are used as resources for improved hygiene behaviors.	20	
2. Positive hygiene behaviors, including correct use and maintenance of facilities, are systematically promoted among staff and schoolchildren.	20	
3. Facilities and resources enable staff and schoolchildren to practice behaviors that control disease transmission in an easy and timely way.	50	

Toilets Sufficient, accessible, private, secure, clean and culturally appropriate toilets are provided for schoolchildren and staff.		
1. Sufficient toilets are available — one per 25 girls and one for female staff; one toilet plus one urinal (or 50 cm of urinal wall) per 50 boys, and one for male staff.	50	
2. Toilets are easily accessible to all, including staff and children with disabilities — no more than 30 m from all users. Male and female toilets are completely separated.	20	
Menstrual Hygiene[1]:		
1. Menstrual hygiene is integrated into the curriculum.	20	
2. Menstrual hygiene-friendly latrines, bathing units/changing rooms, water and hand-washing facilities.	40	
3. Toilets provide privacy and security.	10	
4. Toilets are appropriate to local cultural and social conditions, are age and gender appropriate and accessible for children with disabilities or suffering from chronic diseases (i.e. toilets are child friendly).	10	
5. Toilets are hygienic to use and easy to clean.	20	
6. Toilets have convenient handwashing facilities close by.	50	
7. A cleaning and maintenance routine is in operation and ensures that clean and functioning toilets are available at all times.	20	
Control of insects and parasites disease schoolchildren, staff and visitors are protected from disease spreaders.		

[1] The menstrual hygiene criteria are from *"World Bank. 2017. Providing Sustainable Sanitation Services for All in WASH Interventions through a Menstrual Hygiene Management Approach. Washington, DC: World Bank. © World Bank, https://openknowledge.worldbank.org/handle/10986/26204 License: CC BY 3.0 IGO.*

1. The density of vectors in the school is minimized.	50	
2. Schoolchildren and staff are protected from potentially disease-transmitting vectors.	20	
3. Vectors are prevented from contact with schoolchildren and staff or substances infected with related vector-borne diseases.	10	
Cleaning and waste disposal The school environment is kept clean and safe.		
1. Classrooms and other teaching areas are regularly cleaned, to minimize dust and molds.	10	
2. Outside and inside areas are maintained free of sharp objects and other physical hazards.	50	
3. Solid waste is collected from classrooms, kitchens and offices daily and is disposed of safely.	20	
4. Wastewater is disposed of quickly and safely.	20	
Food storage and preparation:		
1. Food is handled and prepared with utmost cleanliness (hand are washed before preparing food).	50	
2. Contact between raw foodstuffs and cooked food is avoided.	20	
3. Food is cooked thoroughly.	20	
4. Food is kept at safe temperatures.	20	
5. Safe water and safe raw ingredients are used.	20	
Plans and funds for operation, maintenance and replacement		
1. Water system	50	
2. Latrines	50	
3. Hygiene	50	
4. Menstrual hygiene	50	
5. Food	20	
6. Waste disposal	50	
7. Monitoring for continual improvement plan	50	
Totals of 1410, possible and actual:	1410	

Water, sanitation and hygiene standards for schools in low-cost settings (who.int)

Menstrual Hygiene Matters resource book and training guide toolkit are the most comprehensive resources on MHM in developing countries (Wateraid 2012, 2015), pp288-9.

For Rotary, *https://my.rotary.org/en/wash-schools-target-challenge and https://my-cms.rotary.org/en/document/guide-wash-schools.

Rotary One Star: Rotary district governor recognition

Education: Teachers trained to teach hand washing, begin taking daily attendance, and supervise daily group hand washing at mealtimes.

Facilities: Establish low-cost, improved water supply, hand washing facilities, and gender-segregated toilets.

Behavior change: Daily supervised hand washing before school meals and at group toilet times.

Management: Plans created for operational maintenance and menstrual hygiene management.

Rotary Two Stars: Rotary and UNICEF recognition

Education: Teachers trained in hygiene and menstrual hygiene management education. Improved attendance and follow-up with absent students.

Facilities: Safe drinking water is available. WASH conditions meet the pilot country's national standards.

Behavior change: All students use the sanitation facilities at all times and actively participate in tasks to keep the facilities clean.

Management: Menstrual hygiene management plan implemented and training provided to maintain sanitation facilities.

Rotary Three Stars: Rotary and UNICEF recognition

Education: WASH lessons integrated into curriculums and complementary teacher training provided.

Facilities: WASH conditions meet national or World Health Organization standards and regular maintenance is ensured.

Behavior change: Community members involved.

Management: Continuous monitoring and evaluating to check compliance with national standards; data reported to government.

The Target Challenge, Step-by-Step:

Follow these steps to help your club achieve the Target Challenge and improve the lives of students in the community:

1. Know your WASH in Schools country or district-level recognition committee. It is a valuable resource for helping with the Target Challenge in your area.
2. Assess the school's needs and understand its long-term goals. Gather baseline data to help you evaluate your project's measurable outcomes.
3. Submit baseline data, required for recognition. Contact your recognition committee to learn how.
4. Use the information from the school assessment, and collaborate with the community to plan and implement your project's activities.
5. Continuously gather outcome measurements and submit data every six months.
6. Complete and submit recognition applications to your recognition committee.

For Rotary, * and https://my-cms.rotary.org/en/document/guide-wash-schools.

References:

Arole, M.R. and Arole, R. (1994) Jamkhed: A Comprehensive Rural Health Project. Macmillan.

Cairncross, S., Feachem, R. (2018) Environmental Health Engineering in the Tropics: Water, Sanitation and Disease Control, Routledge.

Janoff, S., Weisbord, M., Weisbord, M. R. (2010). Future Search: Getting the Whole System in the Room for Vision, Commitment, and Action. Ukraine: Berrett-Koehler Publishers.

Kelley, A. (2020). Public Health Evaluation and the Social Determinants of Health. Routledge.

Rotary International (2022). Water, Sanitation, and Hygiene Guidelines for Global Grant Funding Document.

Menstrual Hygiene Matters resource book and training guide toolkit are the most comprehensive resources on MHM in developing countries (WaterAid 2012, 2015), pp288-9.

Made in the USA
Monee, IL
03 November 2026

67380404R00046